D0563946

JAVELINAS

RIO NUEVO PUBLISHERS®
P.O. Box 5250, Tucson, Arizona 85703-0250
(520) 623-9558, www.rionuevo.com

Design: Karen Schober, Seattle, Washington

Library of Congress Cataloging-in-Publication Data

Yule, Lauray.
Javelinas / Lauray Yule.
p. cm. -- (Look West)
ISBN 1-887896-61-9 (hardcover)
1. Collared peccary. I. Title. II. Series.
QL737.U59Y85 2004
599.63'4--dc22

2004005797

Printed in Hong Kong
10 9 8 7 6 5 4 3 2 1

JAVELINAS

Lauray Yule

LOOK WEST
SERIES

RIO NUEVO PUBLISHERS
TUCSON, ARIZONA

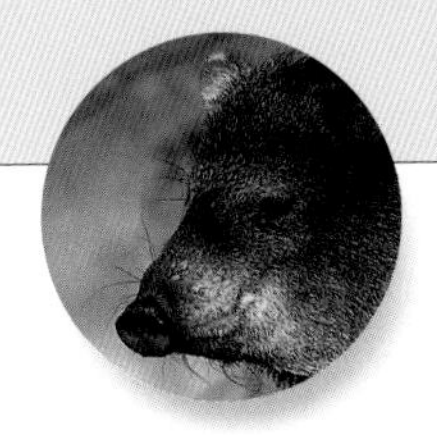

YOU MAY SMELL THEM, BUT NEVER SEE THEM. IF YOU LIVE IN THE DESERT SOUTHWEST—ARIZONA, NEW MEXICO, OR SOUTHERN TEXAS—YOUR YARD MAY HAVE BEEN RANSACKED AND ROOTED DURING THE NIGHT, BUT YOU NEVER HEARD THE INTRUDERS. THESE VISITORS WERE PROBABLY ONE OF THE MOST FASCINATING CRITTERS OF THE NEW WORLD: JAVELINAS.

NO, NOT PIGS

Many people believe javelinas are hairy pigs, but they're not. Some believe they're big rodents—also not true. Vicious predators? Those long canine teeth, honed to a razor edge, may give that impression! Not true, either.

Though pigs and peccaries are classified within the same order of mammals (Artiodactyla, along with deer, antelope, and

"The Collared Peccary," lithograph from The Naturalist's Library *by Sir William Jardine, 1884.*

hippopotamus), they're in different families: *Suidae* (pigs and hogs) and *Tayassuidae* (peccaries). The two families diverged about 38 million years ago: pigs evolved in the Old World, peccaries in the New World.

Only three peccary species exist: javelina (pronounced HAVE-AH-LEE-NAH, also known as the collared peccary), white-lipped peccary, and Chacoan peccary. Both the white-lipped and Chacoan, once thought to be extinct, live only in Mexico and South America.

Of the three types, javelinas are the smallest, as well as the most abundant and most widely distributed: they root and wallow from the desert regions of North America into South America's rainforests and as far south as the woodlands of Paraguay—perhaps the widest distribution of any North American hoofed mammal.

Thinking of the javelina as merely a "desert critter" is just another myth.

ORIGINS

Today's peccaries descended from a gigantic beast that lived about 25 million years ago. Fossil jaws and teeth, found in the Agate Gate Springs Quarry in Nebraska, indicate its skull alone was about three feet long. Modern peccaries are pint-sized in comparison.

The entire peccary family probably evolved in South America. By about a million years ago, the larger ones ranged across the continental United States and may have been the most abundant giant mammal. Ice Age extinctions eliminated many huge mammals that shared the continent, including giant sloths, prehistoric bison, mammoths, and ancestral peccaries. Peccary families in warmer, more temperate, and tropical regions—smaller versions of their big ancestors—roamed the tropical rain and thorn forests, living much as they do today.

The ancient Mayas honored the javelina by naming a goddess the Great White Peccary and a king Precious Peccary.

RAMBLIN' NORTH

Then "wanderlust" struck the white-lipped and collared peccaries. The two species ambled northward on migrations that spread the white-lipped peccary into southern Mexico's tropical forests and the collared peccary into desert regions of the southwestern United States and northern Mexico.

It was not a short-term ramble: javelinas weren't recorded in present-day Arizona until the mid-1700s—less than three hundred years ago. Between 1756 and 1767, Jesuit Ignaz Pfefferkorn and Juan Nentvig both wrote about native people in what is now southern Arizona using "musk hogs" (javelinas) for food.

Early beaver trappers discovered javelinas along streamside bottomlands of the San Pedro and Gila Rivers in 1826. Chronicles from military expeditions thirty years later recorded sightings in these river bottoms as well. Later they were found in Arizona's southern mountain ranges. Records of early sightings in New Mexico are sketchy: naturalist E. A. Mearns saw them along the Mexican border in 1907 while doing a survey for the Smithsonian Institution. Later, in 1931, U.S. Geological Survey teams also recorded small javelina herds along the border. Few accounts exist because explorers, trappers, and prospectors did not venture often into southern New Mexico because of Indian conflicts.

Over the past hundred years, the javelina has become a well-established member of the desert Southwest's fauna scene—thriving and continuing to increase its holdings.

WHAT'S IN A NAME?

The name "javelina" brings to mind myths of fearsome, vicious creatures, living in packs in desert country. Origins for the name are debated: some suggest the word *javelin*—the spear thrown in athletic competitions—is the root, relating to the javelina's spear-like canine teeth. There may be a connection to *jabalina,* a Spanish word for spear—not because of their spear-like teeth, but because spears were used to hunt them. Dictionaries indicate the name comes from the Spanish *jabalí*—a wild boar (a female jabalí is a *jabalina*).

Beaver trappers arriving in Arizona in the 1820s called javelinas "wild hogs." The word "javelina" appeared in letters and articles in Texas in the 1800s, some using the spelling "havalina." By the 1900s in Arizona, the word javelina was, at last, the critter's common name.

PRICKLY, WIRY, BRISTLY

Javelinas have large heads, delicate, slender legs, and relatively small feet. A full-grown collared peccary stands about two feet tall, is about three feet long, and weighs forty to fifty pounds. Boars (males) are generally a little larger than sows (females).

Adults in winter are dark brown to almost black, peppered with white, giving them a grizzled look. Petting an adult javelina would *not* be like cuddling the family cat: their hair is wiry and bristly. Some of these brittle hairs reach six inches in length, forming a dark mane starting between the javelina's ears and running the entire length of the animal.

Very nearsighted, javelinas depend on their senses of hearing and smell.

When a collared peccary is frightened or aroused, it raises this bristly mane, much like a dog raising its hackles. Some early observers thought these bristles might serve as defensive weapons—like porcupine quills. A distinct white collar slightly less than an inch wide runs around the animal's neck, set back almost to the shoulders, more like a cowl than a collar. It's most noticeable in winter.

Newborn piglings (or pig*lets*—take your pick) are reddish-brown or tan, with a dark stripe down the middle of their backs from snout to rump, a trait that carries over to adulthood. Otherwise, they are little one-pound miniatures of their parents. Piglings grow rapidly, and after

Javelinas usually give birth to twins.

three months, they're smaller versions of their herd mates, slightly lighter in color.

COPING WITH SEASONS

Adult javelinas wear their dark winter coats from November to about mid-February. Then, like many critters, they shed and become lighter in color. Some even go bald on their hips and rumps.

Small white patches begin to show on most adults by March and April, and from June to August, all adults are pale—so light that their collars blend in and disappear. Some javelinas are mistaken for albinos at the height of this "molt." The color change is important: it helps the animals cope with high summer heat in the desert Southwest. However, they never quite lose the distinctive mane, which stays dark.

New, darker bristles begin appearing in September, and by mid-October, collared peccaries' coats begin to thicken for winter. Still, javelina "coats" are lightweight, and they wear no underfur. This factor, plus finding suitable habitat, stops javelinas from expanding into new areas of Arizona, New Mexico, Texas and beyond: they can't survive winter in a cold climate.

FOUR, THREE, TWO, ONE

Javelinas differ from pigs and hogs, their distant kin, in ways other than place of origin. Though they appear similar, nuances exist that you might not notice unless you're up close and personal with a javelina:

- Pigs have four toes on each foot. Javelinas have four on their forefeet, but only three on the hind ones. However, both pigs and peccaries walk on only two toes of each foot.
- Javelinas have short, inch-long tails usually hidden under their hair; they also have short, rounded ears that stand upright or fold back when the animal is foraging through thick, thorny vegetation or defending itself. Pigs have distinctive, curly tails and generally "floppy" ears that fold forward over their eyes.

‡ Javelinas have interlocking jaws with thirty-eight teeth; pigs have moveable jaws with forty-four teeth. Javelinas grow straight canine teeth more than two inches long, honed by constant friction to a razor edge; swine canines curve outward.

‡ Hogs produce huge litters of babies after a four-month gestation, but peccaries don't—only one or two little ones are born after a five-month pregnancy. Javelina births happen throughout the year, but most occur in mid-summer or fall, when food is plentiful.

‡ Collared peccaries have four mammary glands (two are functional); pigs have six to twenty.

‡ Internally, a javelina has a complex chambered stomach and no gall bladder; hogs have a simple chamber stomach and a gall bladder.

‡ All peccaries have a scent gland on their backs, four to six inches above the tail. Swine don't have this gland.

HERD LIFE

Javelinas are social animals, traveling in groups of at least three or four—even up to thirty animals, though average group size is about ten. Herds are made up of family members of different ages and sexes. Social structure is strong—continuously strengthened by one-on-one physical contact during feeding, bedding, and play. This develops and reinforces peaceful co-existence within a herd. It's rare to see a lone javelina unless it is old, sick, or injured, and can't keep up with its group.

Peccaries are nearsighted and don't see distant objects, even moving ones. They live in a world of scent and sound. As a javelina herd rambles through its territory foraging for food, there's constant communication with a variety of vocalizations and a group "scent." The scent is a distinct, pungent, "skunky" odor, and serves you almost as well as it serves javelinas: if a herd is near, you'll probably smell it before you see it.

Collared peccaries transfer this scent to one another by pair-ing off, standing side-by-side, nose-to-tail, and rubbing their

heads over the posterior scent gland above each other's tails. Each animal does this with every herd member, sharing its individual odor, creating a unified "blend" for that herd.

Each javelina frequently "freshens up" its herd mates with musk. But musk marking doesn't end with fellow javelinas: rocks, logs, brush, tree trunks—even boundary fences, walls, and buildings within a herd's range—get a pungent "rump-rubbing," meaning: "This is our territory. Trespassers will be run off."

MORE THAN GRUNTS

Vocal contact is a second way javelinas maintain herd integrity. They have a wide sound repertoire—nattering as they forage. Variations in tone, volume, and repetition are important for keeping a group intact, especially in brushy or tall-grass terrain where their poor vision can't help them.

GRUNTS The most common sound they make is a low, soft grunt. It's a contented call and helps a herd stay together. The group does this constantly when feeding. Even little piglings, usually stuck like magnets to their mother (often between her legs, under her belly), grunt almost continuously. Their immature grunt is like kittens' purring. Adult animals make a similar noise that's less repetitive, deeper, and more resonant. This quiet grunting is barely audible unless you're very close to a herd.

Adults also utter a single loud grunt to locate other herd members. Usually the closest group neighbor answers. If a group gets scattered, loud grunts from every animal help them reconvene.

BARKING Javelinas bark—from loud dog-like barks to high-pitched yips—when they're distressed or separated from the herd. Juvenile collared peccaries bark like puppies.

WOOFS Adult javelinas make this loud call if they're startled, spooked, or defensive. If a "woofing" javelina is really upset, it may also react with tooth clacking and squirts from its scent gland. The woof call alerts all herd members. They usually freeze, listen intently, check for scent, and stay alert to run—or attack. If one animal runs away "woofing," other herd members often scatter, too.

Collared peccaries rarely hurt each other during a squabble.

GROWLING Growling is an aggressive call, most often done by adults. It's the loudest sound a collared peccary makes: in ideal conditions, you can hear it more than half a mile away. In a dispute over choice food or bedding areas, the two squabblers stand head-to-head with mouths wide open, displaying their canines, growling fiercely. These encounters are short and don't generally end in fighting.

TOOTH AND JAW POPPING Facing a cornered adult javelina almost always initiates a chilling, intimidating series of machine-gun-like popping sounds as the forty- to fifty-pound animal clicks and clacks its teeth together—especially the two-inch-long canines, longer than canines of any carnivore in North America. Both aggressive and defensive, the display's intensity may amplify with growls, a raised, bristly mane, and squirts of foul-smelling musk.

For less vicious encounters between herd mates, the clacking or chattering intensity and speed increases: sows with young piglings pop their teeth at youngsters or at other javelinas that come too close. Sometimes a feeding collared peccary warns off a herd buddy with a smacking sound made with its mouth and jaws, or gives a single pop.

If a herd encounters another group where territorial boundaries overlap, it turns into a "Hatfield and McCoy" feud: plenty of tooth popping and chasing between clans.

SQUEALING Piglings are the main "squealers" in javelinaland. A baby collared peccary squeals when it's hungry, separated from its mother, or playing. Loud squeals attract one or more adults to investigate, and if necessary, defend the youngster. A high-pitched distress squeal can be heard a half-mile away. Juveniles sometimes squeal when playing or pursued. Adult javelinas seldom squeal unless distressed.

THE DAILY FORAGE

Javelinas generally forage in mornings and evenings, but sometimes also at night, especially in suburban or urban habitats where javelina/human encounters occur.

At lower elevations, javelinas chomp on cactus and succulent plants—particularly prickly pear. Dinner and breakfast may include seeds, fruits, roots and tubers, grasses, broad-leafed flowering plants (forbs), and shrubs. Backyard garden delights such as melons, corn, squash, and fruit, both fallen and those close enough to the ground to pluck, also appear on their menus.

SPINES? WHAT SPINES?

Surprisingly, javelinas eat spiny prickly pear pads with no obvious harm to their mouths, stomachs, or intestinal tracts. Succulents provide both water and nutrition, though javelinas cannot live on prickly pear alone: raw prickly pear is loaded with oxalic acid, a nasty poison that even a peccary's cast-iron digestive system can't handle full-time.

They also enjoy cactus blossoms in spring, to many a cactus gardener's dismay.

IT'S NOT PERFUME

Let's face it: javelinas stink. Though the smell may be sweet perfume to another collared peccary, its pungent odor is repugnant to most humans—and to many other animals.

The source of this essence was a mystery to explorers who saw javelinas from afar but smelled them nonetheless. In 1683, Edward Tyson wrote of the "Mexico Musk Hog," an animal with a "cystis" or scent gland under the skin filled with a yellow "juice." In 1764, Jesuit missionary Father Juan Nentvig described a "navel" in the animals' loin area. He explained that hunters carried hollow reeds, and if they killed a javelina for meat, they quickly inserted a reed into this navel to "vaporize" the musk. Otherwise the meat was tainted and inedible, "regardless of how hungry one may be."

The gland, which actually looks more like a nipple than a navel, secretes an oily, amber-colored fluid that quickly turns black when exposed to air. If a javelina is alarmed or excited, it squirts this oily stuff several inches. They also squirt it on each other when alarmed: if the herd scatters, this fresh dose of "perfume" helps animals regroup.

PREDATOR, NO. OMNIVORE, YES.

Collared peccaries don't turn up their snouts at carrion, bugs, or grubs as they root for underground root morsels. It's a myth that they capture lizards, snakes, and mammals: they're not adapted, equipped, or inclined to do so.

In urban or suburban environments, javelinas acquire a taste for poorly contained garbage (bumping over the cans or containers) or pet food left outdoors. Birdseed, particularly quail blocks, is a gourmet treat. Put one out, and it's devoured overnight.

It's on record that javelinas kill rattlesnakes, but they rarely eat them. Anything you've heard about them gobbling "poisonous serpents" is mythology. Feral hogs and wild boars introduced from the Old World, however, kill and eat rattlesnakes, nonvenomous snakes, lizards, and toads.

Newborn piglings can travel with the herd just a few hours after birth.

MUNCH 'N' CRUNCH

In late summer, when seedpods from trees such as palo verde, mesquite, catclaw, sweet acacia, and ironwood ripen and fall, javelinas enjoy crunchy desert feasts: more than one suburban resident or desert camper has been awakened by what sounds like a fleet of food processors chopping nuts under a window or outside a tent. During winter months, javelinas in desert habitat seek out packrat middens (nests), root them up, then munch on the rodent's legume seed cache.

At higher elevations in pine-oak habitat, collared peccaries may feast on fallen acorns. Javelinas rarely live above six thousand feet in mountain areas: it's too cold, and "good grub is slim pickin's."

FOOD FINDS

Peccaries don't have cutting teeth and can't use their tongues to grasp broad-leafed plants and grasses as many herbivores do. They bite and pull to get a mouthful of greenery, and they chomp off

JAVELINAS AND H_2O

It's debatable how much water javelinas need. Some experts believe they get along for long periods on water from succulent vegetation. Others believe a permanent—or at least semi-stable—watering hole is essential for a healthy herd, especially when sows have babies.

If fresh water is available, javelinas regularly come for a slurp. Free-running streams, creeks, seeping springs, or river potholes and tinajas (permanent or semi-permanent rainpools in hollows in solid rock) within a herd's territory serve as regular stops.

Javelinas approach waterholes together, but the entire herd doesn't go for a drink at once. Some members stay under cover, alert for predators such as mountain lions, bobcats, coyotes—even predatory birds. Herds with small piglets are especially cautious. There's very little squabbling or fussing, and herds don't hang around a water source for long: too risky.

Taking turns at the water hole.

chunks of prickly pear, leaving behind some of the stringy support structure within the cactus pad. Their chief competitors for prickly pear—packrats—gnaw a neat oval or partial circle out of a pad, leaving no untidy fibers.

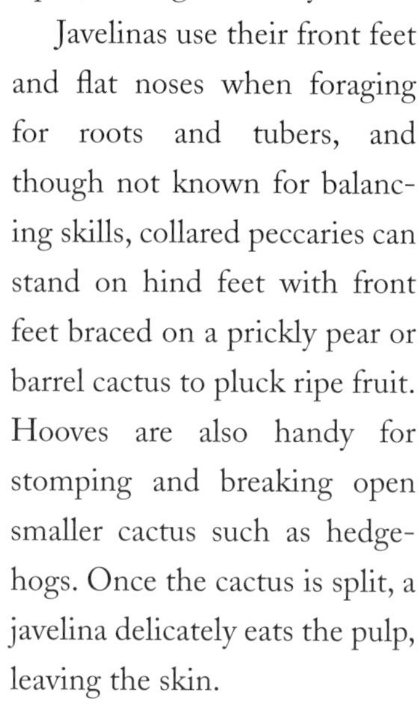

Javelinas use their front feet and flat noses when foraging for roots and tubers, and though not known for balancing skills, collared peccaries can stand on hind feet with front feet braced on a prickly pear or barrel cactus to pluck ripe fruit. Hooves are also handy for stomping and breaking open smaller cactus such as hedgehogs. Once the cactus is split, a javelina delicately eats the pulp, leaving the skin.

Cholla fruit are plucked like prickly pear fruit, or they may be bumped off cacti with some snout shoves, then eaten. Javelinas pull apart yucca and agave plants for the inner pulp, leaving behind heaps of tattered leaves and stalks scattered around gutted plants.

Even two or three javelinas on a hungry stroll can wreak havoc on a landscape.

COOL WALLOWS, WARM COVER

During midday, javelinas find cool, shady places in canyons or washes, or in dense palo verde or mesquite tree stands and brushy areas, and the herd beds down for siesta. Collared peccaries take long naps during hot summer months from about May to September. During fall, winter, and spring, resting times shorten, and you'll see more herd activity during daylight hours.

An average herd's territory is a little over one square mile in size, though it varies, depending on available food and places to bed down or wallow. The animals have up to five bedding areas within this home range. Creatures of habit, they use the same spots year after year and, depending on the season, choose the ones that suit them.

In summer, during high heat, they seek out wallowing holes in streambeds or river bottoms (irrigated areas are fine, too—including a yard) or dense shade. Each javelina scrapes out a resting area with front hooves and snout, several feet away from other herd mates. Bedding down often leads to squabbles as individuals "argue" about choice spots.

During cold winter nights, javelinas "dog pile" or cuddle to share warmth, usually under trees or rock ledges or in caves. Young piglings wedge between herd mates for extra warmth.

PLAY TIME!

Bedtime is playtime—especially for javelina piglings, though juveniles and adults play, too. Youngsters run, leap, and chase back and forth—a peccary version of tag. They flop, flip over, tumble, paw other members of the herd (sometimes met with an annoyed growl), and dash in circles.

Adult and juvenile play is a bit more subdued and resembles mock fighting. There maybe growling, squealing, and head-to-head mouthing—but no biting—mixed in with nose-to-tail spins.

Play sessions occur before and after siesta, and probably reinforce bonds within the herd, though they end quickly: time for a nap—or onward with a daily forage.

HERD HIERARCHY?

Javelina herds aren't rigidly organized. Usually one dominant boar breeds and leads when defending a herd or territory, but non-dominant and juvenile boars aren't driven from the group. Older sows seem to direct daily activities—herd movement, feeding, and bedding—but no sow is a permanent leader, either.

New herds sometimes form and occupy abandoned territories, or territorial boundaries may shift to absorb another group if the animals "agree" to get together. Sows and their acceptance of other "new" herd members play a role in group dynamics. Generally, javelinas are remarkably adaptable and amiable about shuffling herd mates—especially during breeding times.

ONE LITTLE PIGLING, TWO LITTLE PIGLINGS...

Baby collared peccaries are born year-round. However, you'll most often see the one-pound newborns (about the size of cottontail rabbits) trailing their mom like little sticktights in late summer and early fall when there's lots of vegetation for a herd to dine on. Triplet births among javelinas are rare.

Births usually occur at bedding sites, and piglings are up on tiny hooves and following mom within a few hours of entering the world. All herd members defend piglets, and the babies instinctively know, when threatened—whether it's real or imagined—to squeal and run for cover under a larger animal.

TOTALLY TOLERATED

Like all "little kids," baby javelinas love to play. They clamor over their adult herd mates during bedding time. And, as with all grown-ups, this pestering sometimes gets a response—usually a head snap or a short chase to send a youngster off to bother another herd mate.

Piglets nurse full-time for about four to six weeks. They do it often: Mom doesn't have much storage capacity. Other sows producing milk let piglets other than their own have a snack, too. By fourteen to sixteen weeks, they're weaned and foraging full time, though they still hang close to mother.

With all this tender loving care, piglings grow quickly: by the end of its first year, what started as a cute, little reddish-brown piglet weighs about thirty to thirty-five pounds. The juveniles feed and bed down with adults—but still like to tease their elders. They'll even take choice food tidbits away from them. This causes some growling and a semi-aggressive "tug-of-war," but adults usually let a juvenile win.

Javelina boars can breed when they're a year old, sows at about a year and a half. They'll live to age fifteen in the wild, though the average is seven years.

Ceramic javelina by potter Nicolás Ortiz of Mata Ortiz, Mexico, 1997.

JAVELINAS AND HUMANS

TABOOS AND MYTHS For thousands of years, the three peccary species provided an important meat source for most Central and South American tribes, with some exceptions. The Mataco Indians of the Gran Chaco region in Argentina and Paraguay don't eat peccary because they believe they will get toothaches and their teeth may chatter like an angry peccary's. Other tribes in the region believe eating the meat of peccaries or pigs causes ulcers in their noses. Some Brazilian tribes such as the Bororó think peccaries, tapirs, and alligators harbor the souls of dead tribesmen, and they never kill one of these animals unless a magician is handy to exorcise the spirit it's carrying.

Other South American tribes have complicated taboos about eating peccary meat, depending on a person's sex, age, and physical condition. Some tribes won't allow boys approaching puberty to eat it; others believe expectant parents shouldn't consume "musk hog" during the woman's pregnancy, and other tribes tack on another year after the baby is born.

FOLLOW ME! Peccary piglings under three days old readily "imprint" to the first living being they encounter. Piglings born at zoos and "living" museums frequently become imprinted to their keepers, following on their heels, just as they do with their mother and other herd mates.

This tendency to become "tame"—relatively speaking—has made it possible for Indians in the Neotropics to keep peccaries as pets—or even as a future meal.

They are never completely "domesticated," however. There's a difference between true domestication and merely keeping wild animals in captivity.

JAVELINA ALLURE

Perhaps it's because they are one of the largest, most distinctive animals in the desert Southwest that javelinas are intriguing and—in some cases—whimsically endearing. Despite their poor manners, pungent perfume, and a tendency to root up everything from pansies to prized cacti from award-winning landscaping, they are remarkably curious critters that enrich the desert Southwest environment.

So, no Southwest gift shop is complete without *something* "javelina." Little figurines. Plush, cuddly stuffed toys. Metal sculptures, both for indoor and outdoor use. Tiles and trivets. Cards. Kids' books. Photos, lithographs, and paintings. T-shirts, of course. The javelina has somehow managed to trot into our hearts.

HUMAN/JAVELINA ENCOUNTERS

Javelinas adapt well to living around humans. Herds cause landscape, garden, or crop damage by rooting up plants for culinary delights or tearing out irrigation systems to create wallows. They also may decide the cactus in a yard is tastier than anywhere in natural desert and keep coming back for second, third, or fourth helpings.

Javelinas are the stars of a popular children's book.

Though the animals are not normally dangerous, excellent senses of smell and hearing may cause a collared peccary herd to bolt if they get a whiff of you or hear you talking or walking. Since they don't see well, as the herd scatters it's possible to get smacked by one.

"I thought maybe it was a family of bears," says a homeowner in the foothill suburbs of Tucson, Arizona. He'd gone out to confront what he thought was a nighttime intruder. Before he had a chance to shout, he was bowled over by a dark, bulky shape and enveloped in javelina "musk." His landscaping had lured in a herd, which was happily chomping away when he confronted them. In the morning, he found his flowerbeds trashed and a garbage container tipped over: urbanized javelinas on an adventurous foray.

NOT SO SWEET

Baby javelinas are tiny, endearing creatures. However, these one-pound cuties are born "fully loaded," with canine teeth, a functioning musk gland, and an unpredictable temperament. They may look like sweet, furry piglets, *but they are not a domesticated animal.* More than one desert dweller has tried to tame a youngster thought abandoned.

Warning! They're difficult to raise, destructive, and short-fused. What starts out as a cute, tiny baby will one day become a forty- to fifty-pound adult with an attitude—and a javelina's razor-sharp canine teeth inflict a nasty bite.

Urbanized javelinas—those herds or hand-raised individuals that come to expect treats from humans—are the most dangerous, bold, and troublesome. They remain highly territorial and defend what they consider "theirs," whether it is a shady, comfy backyard, a pet food bowl, or a handout: and they *will* bite the hand that feeds them.

Contemporary Oaxacan wood carving by Jacobo Angeles Ojeda.

A WORD OF CAUTION

Observing collared peccaries feeding, cruising from place to place single file, bedding, or visiting a waterhole can be fascinating. If you're quiet and move slowly, it's fairly easy to watch a javelina herd going about their daily lives. Bear in mind: they are easily startled, especially if they smell or hear you.

These are nimble critters who can turn end for end before you even see them do it and can broad-jump six feet from a standstill. They gallop with a stiff-legged gait up to twenty-five miles per hour, sometimes bounding nine feet. Low-slung and tapered, they dash through underbrush, leaving a cloud of dust and a dose of javelina perfume. Water doesn't slow them down: streams and rivers are no obstacles.

Their instinct is to flee by scattering at a gallop—and one or more forty- or fifty-pound animals may head your direction. Watching them from a distance with binoculars is the safest tactic.

Fast and tough, the javelina is the sports mascot for Texas A&M University–Kingsville.

JAVELINAS CAN'T JUMP

Some tips to avoid javelina encounters:

ERECT A FENCE OR WALL AROUND AREAS YOU WANT PROTECTED. Surprisingly, javelinas cannot jump very high. A three- to three-and-a-half-foot fence is sufficient to restrain them, as long as it is sturdy and well-anchored. Javelinas have a tendency to shove and bump obstacles, and they're persistent if they want something on the other side. A fence or wall is particularly important for landscaped areas, gardens, and flowerbeds, which javelinas love.

DO NOT FEED THEM! Javelinas conditioned to handouts are dangerous. Plenty of well-meaning individuals have been bitten by javelinas that wouldn't take "no" for an answer or felt threatened or cornered.

REMOVE PET FOOD AND WATER BOWLS AT NIGHT—even if your pet is contained in a secured pen. The smell of food draws not only javelinas, but other unwanted wild guests.

DON'T PLACE QUAIL BLOCKS ON THE GROUND OR OVERFEED YOUR AVIAN VISITORS. Feed a couple of cups of seed per day: birds eat it during a short feeding session. This eliminates temptation for unwanted javelinas and rodent guests—and it's healthier for birds too, since overfeeding can also lead to the spread of diseases.

SECURE GARBAGE CONTAINERS WITH TIGHT-FITTING LIDS. Urbanized javelinas quickly learn to overturn garbage cans or containers: the scent of improperly bagged and loose garbage is irresistible.

The best advice is to leave javelinas alone. Watch them from a distance. Enjoy wild encounters you may have with these well-adapted, remarkable animals.

ABOUT DOGS

Dogs aggravate javelinas, especially if a herd has newborn piglets. Any dog, large or small, can trigger a defensive attack, even if the dog is leashed. If you encounter javelinas while strolling with your pet, it's best to re-route your walk or retreat slowly and keep the dog from barking. Attacks can result in expensive vet bills—or even the loss of your pet—not to mention some damage to you if you attempt to intervene.

JAVELINA COMPETITORS

COUES' WHITE-TAILED AND DESERT MULE DEER Weighing anywhere from 110 pounds (white-tailed) to as much as 400 pounds (a large mule deer), these cloven-hoofed herbivores share javelina territory, especially at higher elevations. Finicky browsers, deer nibble grasses, broad-leafed plants, and young, tender woody shrubs and trees.

RABBITS AND HARES A desert cottontail may weigh only a pound, but competition comes from sheer bunny numbers that hop through javelina habitat. Jackrabbits (hares)—both black-tailed and antelope jacks—nibble javelina forage. Rabbits and hares are strict herbivores, not omnivores like javelinas.

SQUIRRELS AND PACKRATS Rock squirrels, Harris's antelope squirrels, and round-tailed ground squirrels compete with javelinas for late-summer and fall harvests of cactus fruits; palo verde, ironwood, mesquite, and acacia beans and pods; and acorns at higher elevations. Javelinas raid packrat middens for stored seeds and pods, digging until they find a cache.

JAVELINA PREDATORS

BLACK BEARS The North American black bear competes with javelinas for food and is a primary javelina predator, especially in Arizona and New Mexico mountain ranges. Weighing between one hundred and three hundred pounds, this

Dead cactus becomes a desert "swamp" where javelinas may find moisture and tasty insects.

omnivore rarely ventures to lower elevations: it can't tolerate desert heat. Black bears can kill and eat adult javelinas, though they prefer piglings.

MOUNTAIN LIONS For all three peccary species, mountain lions are primary predators. (In Mexico and South America, jaguars get their share.) Weighing from 100 to 160 pounds, mountain lions are crouch-and-wait predators that ambush deer and javelinas. Collared peccaries sometimes rally and successfully drive off a mountain lion.

BOBCATS This feline omnivore shares almost all javelina habitats, including urban and suburban areas. An adult bobcat weighs twenty to thirty pounds, about half the size of an adult javelina. It attacks collared peccary piglings separated from the herd. Bobcats sometimes hunt in pairs to corner a piglet, but may be driven off by a growling, tooth-clacking adult.

COYOTES It's unusual for a lone coyote to kill a healthy javelina. This well-adapted canid weighs half as much as a mature collared peccary, so piglets and injured or sick adults are targets. Two or three coyotes "team hunting" may trail a javelina group to cull out a straggler or youngster. Javelina herd mates may chase off such wannabe diners.

FOXES Gray foxes share almost all javelina habitats and sometimes snatch a pigling. Weighing only 10 pounds, foxes are omnivorous. They eat fruits, nuts, and berries, as well as carrion—tastes shared by javelinas.

BIRDS OF PREY Eagles and large hawks hunt by day and can kill newborn javelinas. Night-shift birds of prey—great horned and barn owls—also occasionally take baby collared peccaries.

This coyote knows to approach an adult javelina with caution.

JAVELINAS TODAY

The javelina coexisted with humans in its South American range for thousands of years. What makes a collared peccary remarkable is its adaptability: with only three hundred years under its hooves in America's desert Southwest, this critter has rooted out a niche among hostile flora and fauna.

Elusive and bold, strictly social, yet socially flexible, javelinas are complex critters. Perhaps that's why we like them, despite their destructive mischief. Their social order is to be admired and respected. They've adapted well as pioneers in a new country, and it's a sure bet they're here to stay...not unlike us humans.

WHERE YOU MIGHT SEE JAVELINAS

AT THE ZOO

Arizona-Sonora Desert Museum, Tucson, AZ
Audubon Nature Institute, New Orleans, LA
Buffalo Zoological Gardens, Buffalo, NY
Dallas Zoo, Dallas, TX
Detroit Zoological Institute, Royal Oak, MI
Emporia Zoo, Emporia, KS
Fort Worth Zoological Park, Ft. Worth, TX
Living Desert State Park, Carlsbad, NM
The Living Desert Zoo and Gardens, Palm Desert, CA
Omaha's Henry Doorly Zoo, Omaha, NE
Orange County Zoo, Orange, CA
Palm Beach Zoo at Dreher Park, West Palm Beach, FL
Phoenix Zoo, Phoenix, AZ
Roger Williams Park Zoo, Providence, RI
Rosamond Gifford Zoo at Burnet Park, Syracuse, NY
San Diego Zoo, San Diego, CA
Santa Fe Teaching Zoo, Gainesville, FL
Smithsonian National Zoological Park, Washington, D.C.
Tulsa Zoological Park, Tulsa, OK

IN THE WILD

Javelinas are also often seen in state and national parks throughout the southwestern United States, including:

Big Bend National Park, TX
Chiricahua National Monument, AZ
Guadalupe Mountains National Park, TX
Organ Pipe Cactus National Monument, AZ
Sabino Canyon Recreation Area, Tucson, AZ
Saguaro National Park, Tucson, AZ

SUGGESTED READING

California Center for Wildlife, with Diana Landau and Shelley Stump. *Living with Wildlife: How to Enjoy, Cope with, and Protect North America's Wild Creatures around Your Home and Theirs.* San Francisco, California: Sierra Club Books, 1994.

Schnoeker-Schorb, Yvette, and Terri L. Shorb (editors). *Javelina Place: The Controversial Face of the Collared Peccary.* Prescott, Arizona: Native West Press, 1999.

Sowls, Lyle K. *Javelinas and Other Peccaries.* College Station, Texas: Texas A&M University Press, 1997.

RECOMMENDED WEBSITES

http://animaldiversity.ummz.umich.edu
www.desertusa.com/magnov97/nov_pap/du_collpecc.html
www.livingdesert.org/sgjavelina.html
www.nsrl.ttu.edu/tmot1/tayataja.htm

PHOTOGRAPHY © AS FOLLOWS:

Erwin and Peggy Bauer: page 47
Paul and Joyce Berquist: pages 5, 40, 51, 54, 57, 58, back cover
Larry Ditto/KAC Productions: page 4
garykramer.net: page 28
Greg W. Lasley/KAC Productions: page 33
Tom and Pat Leeson: pages 14, 23
McDonald Wildlife Photography: page 24
Wyman Meinzer: pages 1, 27, 49, 61
Richard Rowe: page 46
Joel Sartore: pages 18, 20
Tom Vezo: front cover
Marjorie Wagner/WildNaturePhotos: pages 9, 31
Jeremy Woodhouse: pages 3, 13, 34, 36

Image on page 41 comes from *The Many Faces of Mata Ortiz,* Susan Lowell et al., Rio Nuevo Publishers, 1999 (photo by Robin Stancliff). *The Three Little Javelinas* book shown on page 44 (text © 1992 by Susan Lowell, illustration © 1992 by Jim Harris) used by permission of Rising Moon.